Children's Science Library

TRANSPORT

Author & Illustrator

A.H. Hashmi

Editor

Rajiv Garg

Published by:

F-2/16, Ansari road, Daryaganj, New Delhi-110002
☎ 23240026, 23240027 • *Fax:* 011-23240028
Email: info@vspublishers.com • *Website:* www.vspublishers.com

Branch : Hyderabad
5-1-707/1, Brij Bhawan (Beside Central Bank of India Lane)
Bank Street, Koti Hyderabad - 500 095
☎ 040-24737290
E-mail: vspublishershyd@gmail.com

Distributors :

- **Pustak Mahal®**, Delhi
  J-3/16, Daryaganj, New Delhi-110002
  ☎ 23276539, 23272783, 23272784 • *Fax:* 011-23260518
  E-mail: sales@pustakmahal.com • *Website:* www.pustakmahal.com
  Bengaluru: ☎ 080-22234025 • *Telefax:* 080-22240209
  Patna: ☎ 0612-3294193 • *Telefax:* 0612-2302719
- **PM Publications**
  - 10-B, Netaji Subhash Marg, Daryaganj, New Delhi-110002
    ☎ 23268292, 23268293, 23279900 • *Fax:* 011-23280567
    E-mail: pmpublications@gmail.com
  - 6686, Khari Baoli, Delhi-110006
    ☎ 23944314, 23911979
- **Unicorn Books**
  Mumbai :
  23-25, Zaoba Wadi (Opp. VIP Showroom), Thakurdwar, Mumbai-400002
  ☎ 022-22010941 • *Telefax:* 022-22053387

ISBN 978-93-814484-0-3
Edition 2011

Printed at : Param Offset Okhla, Delhi

CONTENTS

747
620.
103

INVENTION OF WHEEL

The wheel has played a very important role in the development of human civilisation. In fact, the wheel has become the pivot of the whole human civilisation. If the wheel had not been invented, there would have been no car, no scooter, no cycle, no train and no aeroplane. The wheel has marked the beginning of scientific and industrial age. Almost all machines make use of the wheel in some form or the other.

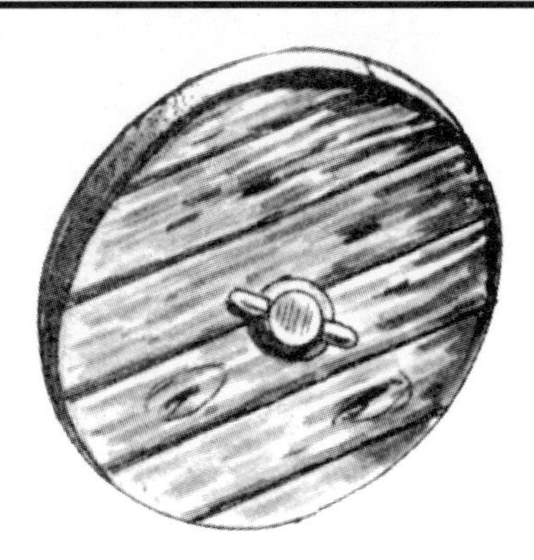

Exactly when and where the wheel was invented is not known to us. Man did not have any means of transport about 15,000 years ago. At that time, man's biggest worry was food and shelter. He used to go into the forest and kill some animal for food using his primitive weapons. After killing a large animal, he had no means of transport to bring that animal to his cave. He used to drag or lift the animal to his cave. It must have been a very tiring job for him.

It is said that necessity is the mother of invention. He might have put some wooden planks under the hunted animal and would have made his first sledge. This would have proved a better way of bringing the large animals to the cave and, perhaps, this was his first means of transport. After this, a modified sledge was developed by putting leather belts between the two wooden planks. This type of sledge proved very useful in snow-bound areas.

It is believed that the Pyramids of Egypt and Stonehenge were constructed by using rollers for transportation. When were the rollers for transportation invented is not known. It is supposed that the idea of rollers might have come from the round stem of a tree with which carrying of rock pieces would have been easier for the ancient man. He might have started using rollers of tree stems to carry heavy things to the desired places.

The sledge-cart of primitive man: the first means of transport

The idea of wheels might have come from the round stem of a tree

Wooden rollers were used by man as a means of transport during the middle Stone Age. After a long span of time, perhaps, the man might have cut a disc from the stem of a tree and made a hole in its centre. He might have used it as a primitive wheel. After putting two such wheels on one axle and a wooden plank on it, he might have used it as the first cart. In this cart, both the wheels and axle rotated together. After this, the potter's wheel was invented.

According to some evidences, the wheel was invented in Syria and Sumeria in about 4000-3000 BC. Around 3000 BC, the wheel was being used in Mesopotamia. In about 2500 BC, the wheel came to be used in the Indus valley also.

For a long time, the wheel remained crudely shaped, because it was made with the help of stone tools. Only after the development of metal tools, a better shape to the wheel could be given.

The Pyramids of Egypt: a miraculous example of the use of rollers for transport

Egyptians could know about the wheel after several hundred years. Spoked wheels were invented in about 1800 BC in Egypt. The

Transportation through rollers

Egyptians made the spoked wheel by joining its inner and outer circles with small bars i.e. spokes. Instead of placing a wooden plank on the wheels, the Egyptians made a box-like cart of it. One or two persons could sit in this cart with some of their belongings.

This two-wheeled carriage was also adopted by the Greeks and the Romans. In the beginning, bullocks were used to pull the cart. Later on, horses were used for the same purpose. The Romans started using a four-wheeled cart which was called a chariot. With the passage of time, many improvements were made in these carts and wheels.

The wheeled carriage marked the beginning of road transport. This gave birth to bullock carts, horse carts, buses, cars, trucks, scooters, bicycles etc. which are being used all over the world for transportation purposes. Roads were first constructed during the Roman empire, but with

Stonehenge: The famous monument of Britain

The wheel was being used in Mesopotamia in about 3000 BC.

its fall, the condition of these roads worsened. The industrial revolution in 18th century introduced the cemented and coaltar roads. Germany was the first country to construct national highways before the Second World War. Hitler made about 3200 km long roads in Germany. At present, USA has the longest network of roads with the length of 6.4 million kilometres. India has a 16,75,000-kilometre long network of roads. About 40% of the total load is carried from one place to another by roads.

The main roads which connect a country from one end to another are called 'highways'. The roads which connect the capitals of two countries are known as 'international highways'.

The road that connects Kolkata (in India) to Peshawar (in Pakistan) is the oldest highway of our country, which is functional even today. The biggest highway in the world is the Pan-American Highway.

■■

BICYCLE

The bicycle is a popular vehicle of our time. It can be seen almost in every house. The shortage of petrol has made it even more important.

A modern bicycle consists of two wheels fitted in a frame. The rear wheel is rotated on a chain and a toothed wheel with the help of pedals. A handle turns the bicycle in a desired direction. Karl Von Drais invented the bicycle in 1818. It consisted of two wheels fitted in a wooden frame and a seat in the centre for sitting.

One day, people saw Drais driving his bicycle on the roads of Mahaim in Germany. He wore a green coat and a high hat and was pushing the bicycle by his feet. The ride on this strange vehicle became a subject of laughter for the people.

Drais named his bicycle as Draisine. The invention of this jocular vehicle by a high official like Drais made his senior officers unhappy and, as a result, his services were terminated. Instead of praising his invention, everyone criticised his bicycle. He died in poverty in 1851.

Later on, Draisine was further improved in England, France and USA. After 20 years i.e. in 1839, Kirkpatrick Macmillan, a blacksmith in Courthiel, invented a lever-driven bicycle. It was named as velocipede. After 10 years, a German mechanic, Philip Heinrich Fisher, fitted a paddle in the front wheel of Macmillan's bicycle. In 1876, J. Lawson fitted a toothed wheel and paddles between the two wheels. Swis Hans Ronald invented the roller chain. After this, wheels with wire spokes, springed seat, ball bearing, gear, gear shift and freewheel were fitted in the bicycle.

In 1874, James Starley of England and H.J. Lawson developed the modern bicycle. There are about 24 to 40 wires in its tyres that work as a spoke. They can tolerate the jerks of the uneven roads. To make the bicycle lighter and faster, John Dunlop invented pneumatic tyres. This invention made the bicycle a popular vehicle. In 1902, the Sturmey-Archer Company of Britain established the speed gear system in the rear wheel hub.

In 1905, India started to import bicycles from England. India started making its own bicycles in 1938. In China, about 80 million people use bicycles today. Every year a cycle-racing competition is organised in France, which is called 'Tour de France'. It was started in 1903. This race is completed in 20 days and covers 4,000 kms.

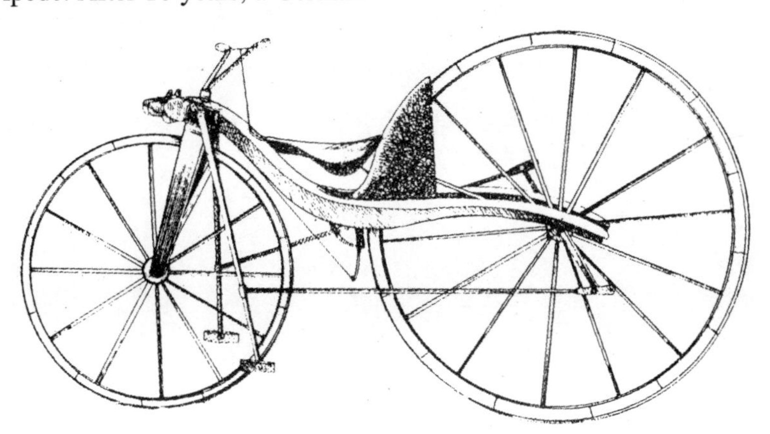

The Bicycle of Karl Von Draise

RAILWAYS

Nowadays, trains are being extensively used for carrying goods and passengers from one place to another. The real development of railways took place during the last 150 years only. In fact, the railways is an outcome of industrial revolution. The spread of railways is linked with the power of steam. The first steam engine was made in 1776 by James Watt of Britain. After this, many modifications and improvements were made in the steam engines, which led to the development of railways.

The first automatic steam locomotive was made in 1803 by Richard Trevithic. The steam locomotive could not move successfully on roads, and it was concluded that steam locomotive cannot travel on roads. That is why, the steam locomotive was run on rails.

James Watt (1736-1800)

The first mechanically propelled vehicle was 'The Rocket', designed by a Britisher named George Stephenson in 1814. He became famous all over the world as the inventor of railway engines. He was born in 1781 in a village near New Castle. He was very fond of engines right from his childhood. He did not study upto the age of 18, after which he joined a night school, where he could get some education. He used to work in a coal-mine. There he was known as the 'Engine Doctor' because he knew about engines more than even the qualified engineers. His master was very happy with him. He helped Stephenson in the development of steam engines.

After two years' hard work, the rail engine, named Blucher, was made. It was capable of pulling 8 carriages, loaded with 30 tons of coal, with a speed of 7km/hr. This engine gave Stephenson confidence that trains could be used in future as a public transport system.

Stephenson presented a plan of a rail line before the Westminsters Parliament house. The Parliament house, with a great difficulty, sanctioned a rail line, from Stockton to Darlington. This rail line was meant to carry both goods and passengers. Stephenson was given the contract of making rail engines.

On September 27, 1825, a ten-mile long railway line was inaugurated. The engine was named 'Active'. Stephenson himself was its driver. This train consisted of 33 bogies, out of which 4 bogies were loaded with coal and flour,

George Stephenson (1781-1848)

In 1829, George Stephenson's 'Rocket' won a reward of 500 pounds

one bogy was for the people of the company, 22 bogies for general passengers and 6 bogies at the end were also loaded with coal. This train with 600 passengers travelled a distance of 8 kilometers in one hour.

In 1826, when another proposal was sanctioned, Stephenson made his second engine for the railway line between Liverpool and Manchester. At that time the government took a policy decision and declared that other builders of railway engines should also get a chance of making steam engines. To give the contract of making railway engines, a competition was arranged with a declaration that whichever engine will be the most successful, the government will purchase it for 500 pounds. In this competition, four engines took part.

First of all ran Stephenson's engine 'The Rocket'. Other three engines could not compete with it and 'The Rocket' was declared successful. This engine could pull a load of 13 tons for a distance of 19 kilometers in 65 minutes. This exercise was repeated 20 times. Seven such engines made by Stephenson were used to inaugurate Liverpool-Manchester line on September 15, 1830.

Stephenson's efforts made trains a grand success. This made him a rich person. He died in 1848.

A railway line was laid from Liverpool to Manchester

Soon after England, railway lines were laid in other countries also. In 1863, the first underground railway was made in England. In 1877, vacuum brakes were introduced in trains which reduced the number of accidents to a great extent.

The diesel and electric locomotives are now gradually replacing the steam engines

After this, many new revolutions came in the train industry. In 1884, electric locomotives were introduced which reduced air pollution to a great extent. In 1925, diesel locomotives were developed in Canada. In 1964, bullet trains were introduced between Tokyo and Osaka in Japan. This train had a speed of 163 km/hr. By 1976, a fast diesel train having a speed of 230 km/hr was already in use in Great Britain. After this, Meglev train, monorail etc. were also developed.

The first train in India was started between Bombay and Thane on April 16, 1853. The biggest railway line in the world is between Moscow and Vladivostok. This railway line is 9,438 kms long. Its name is Soviet Trans-Siberian Line. Development of modern technology has made it possible to manufacture tube trains, channel trains and trains without drivers. Now, trains having a speed of more than 270 kms per hour have also been manufactured.

■■

SHIPS

Most of the ancient civilisations emerged at the banks of the rivers. At that time, crossing the rivers was a big problem to man. When and how did the idea of making a boat come in the mind of man is not known but the primitive boat would have been a log of wood. He might have made the first boat by hollowing a log of wood. After that, he would have made the front part slightly sharp and made the real boat.

According to known facts, perhaps boat-making started in about 4000 B.C. Masts and sails were developed around 3500 B.C. In the beginning, the use of boats was confined to Dazala, Farat and Nile rivers. Egyptians were the first who used boats in the sea. The oldest boat which could be found and is safe even today was constructed in about 2500 B.C. This boat is 43.4 metres long and weighs about 40 tons. It was found buried in Egypt near the Khufu Pyramid. The oldest wrecked ship is assumed

The idea of making a boat might have originated by seeing a floating log of wood

A Phoenician mast-boat

A Portuguese warship

to have been constructed in about 2700 to 2250 B.C.

Ancient Phoenicians used to make trips around Africa by boats. These were the first Mediterraneans who traded with Britain by sea and identified sea-routes to India in about 600 B.C. Harbours were also made by them. Egyptians, Greeks, Romans, Indians and Arabs carried out trade on a large scale with each other. Greeks, Romans and Vikings also made ships. Vikings used to call them as long ships.

After the invention of magnetic compass, bigger ships were made. The ship-making technology could be fully developed only in the 14th century. In this field Italy was the first, and then the lead was taken by England. The Portuguese and Spanish were their rivals. For about 100 years, mast ships could not replace sail ships. These were developed by the Britishers along with their naval forces. In 1620, the submarine was developed. In the 17th and the 18th centuries, sailing ships acquired larger speeds and became more capable.

In the 18th century started the construction of steam-driven ships. Britain was at the top in making such ships. A ship named 'East India Main' was constructed in England. For faster movement 'Frigate' boats were made. The Cliper ship was also developed which was used to carry tea from South-East Asia to England. By the year 1863, wood was replaced by iron in the ship-building industry.

The first steam-driven Pyroscaphe steamer was made in 1783. The first successful steam-driven ship Charlotte Dundas was made in 1801-1802. Paddle and wheels were fitted in its rear part. Screw propeller was invented in 1836.

An Aircraft Carrier Enterprise

Bartha is the oldest mechanical ship which was 50 ft. long and its capacity was 48 tons. It was designed in 1844.

A British engineer, Charles Parsons carried out the first experiment of steam turbine in a ship. Nine diesel engines were used in 1911 in ship building. The American submarine, Nautilus, driven by nuclear power, was made in 1955. Today there are many submarines that are driven by nuclear power and they are capable of staying inside the water for months. The first nuclear-powered ship named Sevana was made in U.S.A. in 1961.

In modern ships the size of the hull, engine, propeller, rudder etc. is very large. In ships, today, we have the facilities of electric generation, bedrooms, bathrooms, kitchens, recreation etc. These ships have turbine engines.

A Missile Launching Cruiser

A ship floats in water because the weight of the ship is less than the weight of the water displaced by it. For stabilising the motion of the ship, stabilisers are used. Ships may be several hundred meters long and can carry 100 millions of tons of goods. They can carry upto 2500 passengers.

Today, we have different kinds of ships for different purposes, such as, bulk carriers, oil tankers, container ships, dry cargo ships, lighters, oceanographic research ships, fish processing ships, war ships, petrol ships, frigates, destroyers, passenger ships, aircraft carriers etc.

Sonars, earthquake-detecting and ice-breaking equipment are fitted in the modern ships. The ice-breaking Russian ship, Sibir, is driven by nuclear power. It is of 75,000-horse power and has the capability of breaking 4-metre thick ice.

Some super tankers are as big as 450 metres in length and they can carry 5,00,000 tonnes of crude oil at a time. In 1960, a bathescaphe was constructed. It was named Troste. It could go 10,917 metres deep into the sea.

The American-built aircraft-carrying ship, Enterprise is 335 metres long. Eight nuclear engines are fitted into this ship.

A submarine driven through nuclear power can carry 16 directional missiles, each fitted with 10 war-heads.

The ship named Elvin was used to take out the remains of the Titanic ship. It went 3,800 metres deep into the sea.

■■

A Submarine

HOVERCRAFT

A hovercraft or aircushion vehicle is such an amazing means of transport that hovers on a cushion of air. It can travel over any fairly flat surface. In this vehicle, a fan sends air into the pipes fitted at its bottom. This air comes out from the holes and makes a cushion of air. A hovercraft can move on land, water, marshy places, snow etc.

The credit of developing the hovercraft goes to a British engineer Sir Christopher Cockerell. The idea of developing a hovercraft came in his mind in 1954. He got his idea patented on December 12, 1955. His idea was that a lot of power is wasted in a ship, which can be saved by making it travel on the surface of water on a cushion of air. It will not only save energy, but larger speeds can also be attained in less power. For this purpose he carried out many experiments.

In a hovercraft, a fan feeds air to the underside of the craft and the openings at the sides provide for the escape of air currents. The air is kept inside the cushion by a plenum. This allows the vehicle to hover two metres above the water or snow or land.

When the first trial of a hovercraft was carried on May 30, 1959 at Cowes in England, it impressed people so much that its news spread far and wide. When a hovercraft of 4 ton weight travelling over the surface of water came to the bank and from there on to the ground, people were very astonished to see it.

If a big-sized hovercraft is made and allowed to travel at larger speeds on the waves of water, the passenger does not feel the pitch and the rolling motion. A hovercraft can carry passengers and freight between ports. A hovercraft can travel with a speed of 150km/hr.

After the invention of a hovercraft by Cockerell, many modifications have been made in it. The first public service of the hovercraft began in July 1962 by V.A. 3, with a capacity for 24 passengers. It was able to travel with a speed of 60 knots (111 km/hr.). This service was across Dee Estuary. The biggest hovercraft of the world was made in Britain which weighs 305 tons and is 56.69 metres long. Its name is SRN 4 MK III. It can accommodate 418 passengers and 60 cars. Its maximum speed can go upto 65 knots (120.5 km/hr). Presently, there are many hovercrafts of all sizes. These are known as ferries. Some of the hovercrafts are so big that even cars, buses, trucks etc. can also go straight into them.

A Hovercraft

MOTOR CAR

A motor car is a petrol- or diesel-driven four-wheeled passenger vehicle. As a road transport, it is very popular among people. Its development is a result of continuous efforts of the scientists.

In 1769, Nicolas-Joseph Cugnot developed two steam-driven tractors. The first one of these two was the first passenger vehicle. Its speed was 3.6 km/hr. The first steam-driven motor car in Britain was developed by Richard Trevithick in 1801. The internal combustion engine was invented in 1876 by the German engineer, August Nikolaus Otto. The petrol-driven car

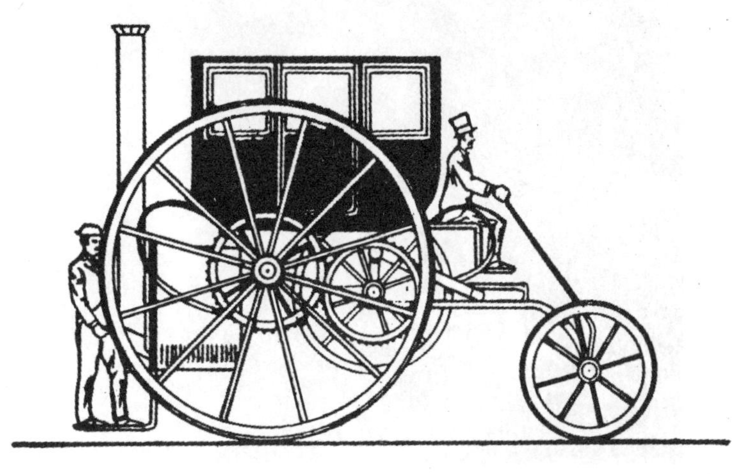

The first successful steam-driven motor car was made by Richard Trevithick

Karl Beng (1844-1929)

Karl Beng made the first successful car with internal combustion engine

The use of petrol in passenger vehicles was started in about 1885

The model T Ford 'Tin Lizzie' (1908-1927)

The Lanchester, 1895

with internal combustion engine was called a 'Motor Wagon'. It was developed by Karl Friedrich Beng of Karlsruhe in 1885. This three-wheeled vehicle weighed 250 kg and its speed was 13 to 16 km/hr. Its single cylinder four-stroke engine produced 0.85 horsepower by 200 rotations per minute. In 1885, he made two more vehicles, one of them is kept still in working condition at the Deutsches Museum, Munich.

Karl Beng, a German engineer, made a three-wheeled car in his small workshop by using the internal combustion engine of Otto. It had a four-stroke petrol-driven engine. Beng developed electric ignition system for this car. He also used water to keep the car cool. It had a seat for the driver and the passenger. A small wheel was fitted on a stick as a steering wheel to change the direction of the car. This car, developed by Karl Beng, was exhibited in an exhibition in Paris in 1887, but nobody gave any attention to this car. Beng started driving his car in Munich. People became astonished seeing his car and he started getting orders for motor cars. In fact, Karl Beng marked the beginning of the modern car industry.

The motor car industry flourished very fast. Henry Ford of USA manufactured his first car in 1896 in Detroit. This car had a four horsepower engine with two cylinders. Henry Ford had the capability of improving the already invented cars. He could understand the shortcomings of the European cars. He developed a low cost, strong and stable car. His model T Ford or Tin Lizzie became so popular all over the world that he became famous and very rich. He opened a new car factory in Detroit, which was 300-metre long. He made assembly lines in his factory so that several cars could be made at a time. From 1908 to 1927, about 18 million Tin Lizzie cars were sold. After this, Ford made several beautiful models of cars. Even today, the Ford cars are too expensive.

The Rolls Royce, 1905

The Mercedes Sports Car, 1928

The Union C Car, 1936

Today, several kinds of cars are being made in different factories all over the world. C.S. Rolls and Henry Royce made costly cars. After this, a four-seater baby car named Austen Saven was made, which was liked by the people.

A modern car has an engine, transmission, drivetrain, steering, brake, suspension, chasis etc. as its parts. All cars have either petrol or diesel engines. When a car is started, a spark of fire from the spark plug burns the fuel. The hot gases produced by the fuel make the car move. Today, scientists have developed computer-controlled and noise-controlled cars. Cars have become an essential part of our lives. Every year, car production goes up by 15%. The motor car, in a short span of time, has become a very successful transport system for us. Presently, many big cities of India are flooded with cars. Cars fitted with A.C. are favourite amongst many people. Jeeps are available for comfortable driving on uneven roads. Sports cars and racing cars are also available in the automobile markets. Computers have played a very vital role in the modern car technology.

The Penhard, 1892

The Duria Racing Model, 1895

MOTOR CYCLE

A motor cycle is a two-wheeled petrol-driven means of transport. The first motor cycle with an internal combustion engine was made by Gottlieb Daimler of Germany in 1885. Its frame was made of wood, and its speed was 19 km/hr. In 1898, two motor cycles were made in Britain. These were named as Holden Flat-Fore and Clide Singal.

Motor cycle industry developed only in the beginning of the present century. The first large scale manufacturing factory of motor cycles was established in West Germany at Munich by Heinrich, Wilhelm Hildebrend and Alois Wolfmuller. In this factory, more than 1,000 motor cycles were made in the first two years. All these motor cycles had four-stroke, two-cylinder, water-cooled, 1488 cc engines. They were capable of producing 2.5 brake horsepower (bhp) by 600 rotations per minute. Mopeds are cheaper and are capable of moving up on elevated roads. Their engines are of the capacity of 50 c.c. or less and have a speed limit of 50 km/hr. The engine of the scooter is also small in size.

A modern motor cycle is fitted with a two or four-stroke internal combustion engine. Some motor cycles also have rotary wenkel-type engines. These engines are either air-cooled or water-cooled. In an internal combustion engine, fuel mixed with air is burnt by the spark produced by a spark plug. The gases produced by the burning of the fuel push the piston in the cylinder. The piston is connected to the wheel. This makes the motor cycle move. The frames of modern motor cycles are made of steel.

Today, several kinds of motor cycles are available in the market. Motor cycles for general use can attain a speed upto 248 km/hr while those meant for racing can attain a speed of more than 300 km/hr. Their engine's capacity is 125 c.c. The engine capacity of big cars can be upto 1,000 c.c.

A motor cycle

AIRCRAFT

Man has been fascinated from ancient times by seeing the birds flying in the sky. To give a practical shape to this fascination, Leonardo Da Vinci made the design of a flying machine and parachute. In the 18th century, two brothers Joseph Michel and Jacques Etienne Montgolfier, made man fly in air with a hot air balloon.

They made a silk balloon and suspended a cage from it. In this, they sent a chicken, duck and sheep first. Hot air was supplied from below

The first balloon in which two men took a flight, was made by two brothers, Joseph Michel and Jacques Etienne

Wilbur Wright and Orville Wright

by straw fires and people watched it rise in the air. It stayed in the air for about 8 minutes.

Finally on November 21, 1783, two young Frenchmen, Pilatre de Rozier and Marquis d' Arlandes rose in a Montgolfier balloon and made a free flight for 25 minutes over Paris. They were the first human beings to fly.

The practice of inflating balloons with hot air soon came to an end and people started using hydrogen gas in balloon. After balloons, Wilbur Wright and Orville Wright of USA marked the beginning of gliders. These brothers were bicycle makers in Dayton, Ohio. In their gliders, the rear edges of the wings were made flexible and were worked by ropes in order to control the lateral or side-by-side position. The gliders of the Wright brothers proved to be very successful.

After several years of hard work, the Wright brothers succeeded in 1903 in making a powered plane, fitted with a petrol engine of 12 horsepower under the right side wing. On the left side, there was a seat for the pilot. The engine was connected with two propellers fitted at the rear of the aeroplane.

On December 17, 1903 at 10.35 a.m. at Kitty Hawk, North Carolina, Oriville got into his plane named Flyer. The motor was started and warmed

up while his assistants held the plane in position. The signal was given and the craft was released. The truck slid down the rail. The plane rose in the air, leaving the truck behind. It flew for a distance of 36.6 metres, remaining aloft for 12 seconds with a speed of 48 km/hr and then landed on its skids. It was the first powered flight in world history. This was a two-winged plane. This flight was witnessed by Orville's brother,

Wright Brothers made a demonstration flight

four others and one child. This plane is still lying safe in National Air and Space Museum at Washington D.C.

After this success, Orville Wright made several improvements in the plane and had a demonstration flight in 1908 in France. After a year of the first successful flight in 1903, Wilbur Wright died. Orville saw the development stages of aeroplanes. He lived upto 1948. By that time, aeroplanes capable of flying with a speed of 100 km/hr were already in use.

In a short span of time, many countries developed the technology of making airplanes. In 1909, Louis Bleriot of France crossed the English Channel. Two Britishers, Alcock and Brown, made a flight across the Atlantic Ocean

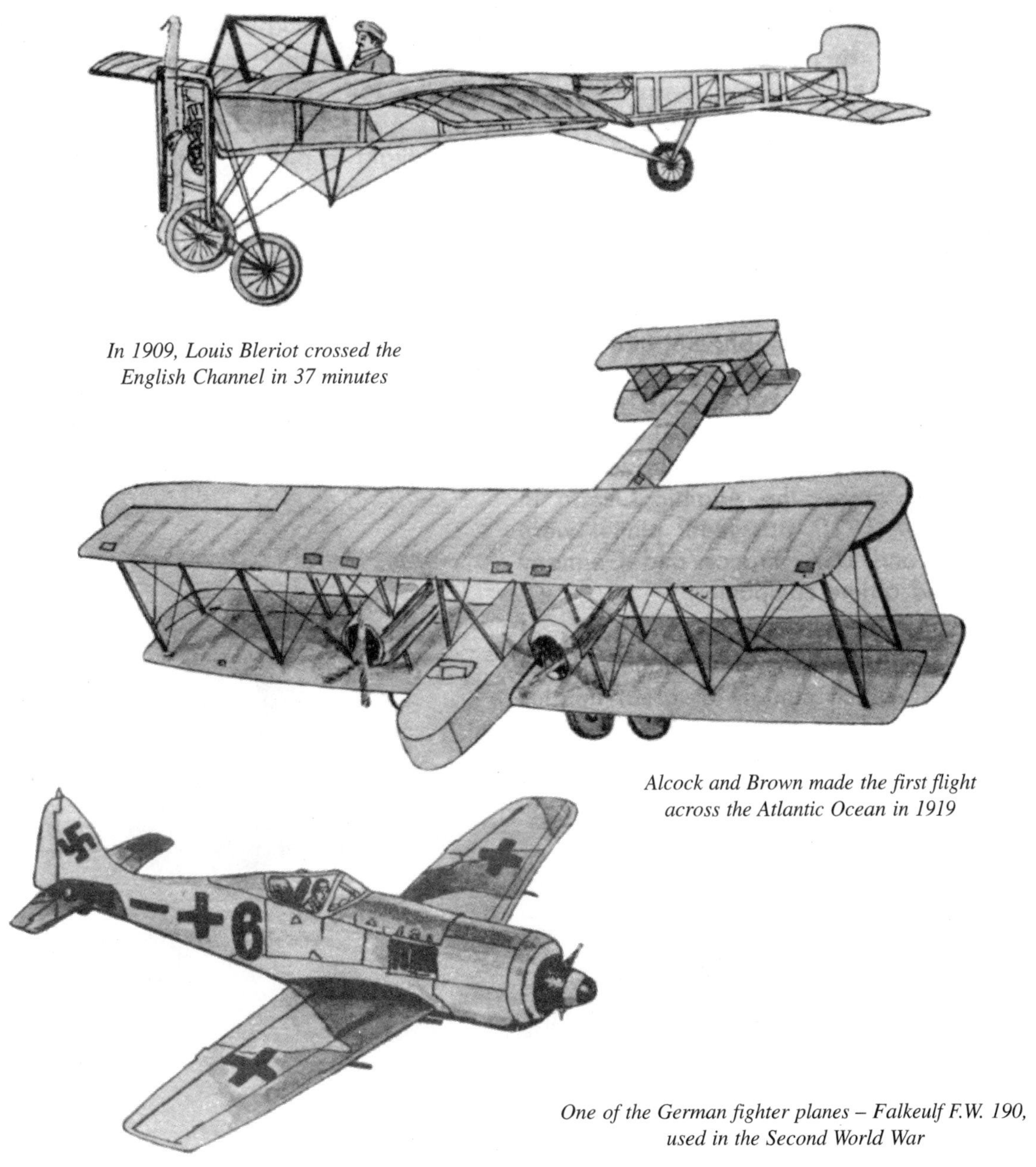

In 1909, Louis Bleriot crossed the English Channel in 37 minutes

Alcock and Brown made the first flight across the Atlantic Ocean in 1919

One of the German fighter planes – Falkeulf F.W. 190, used in the Second World War

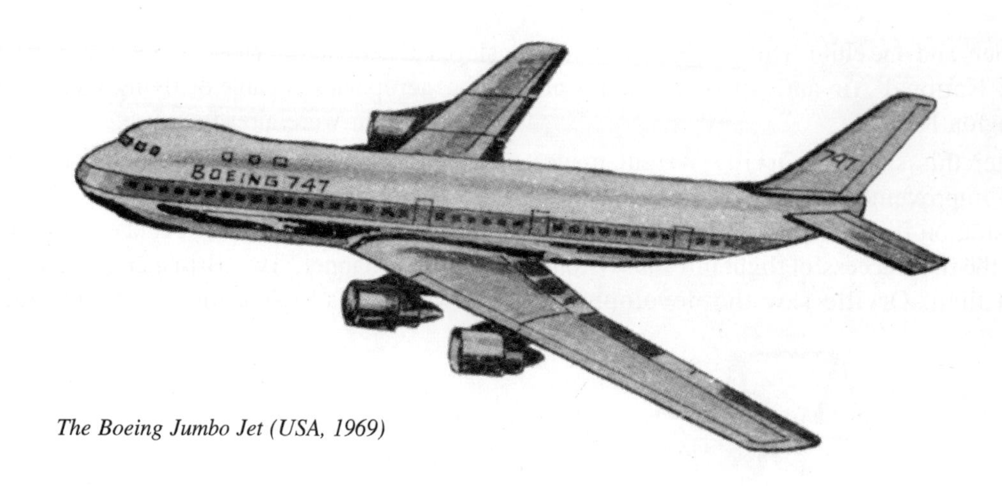

The Boeing Jumbo Jet (USA, 1969)

in 1919. In 1920, passenger planes were made and USA, France and Germany opened airline companies.

Today, we have several types of planes. The most modern planes are jet planes, which can fly at great heights with speeds faster than the speed of sound. These planes mainly have three parts–body, wings and rear part assembly. The body has a cockpit for the pilot and seating arrangement for the passengers. At the tail, there is a rudder. The rear part gives the stability to the plane. During flight, a plane is under the action of four forces, namely, weight, lift, thrust, and drag. These four forces keep the plane in the air and make it fly.

Today, aircrafts having the speed more than that of sound have also been manufactured. There are also such aircrafts that can take off straight and even land in the same way, thus not requiring a runway for taking off and landing on the ground. The shape of their wings can also be changed.

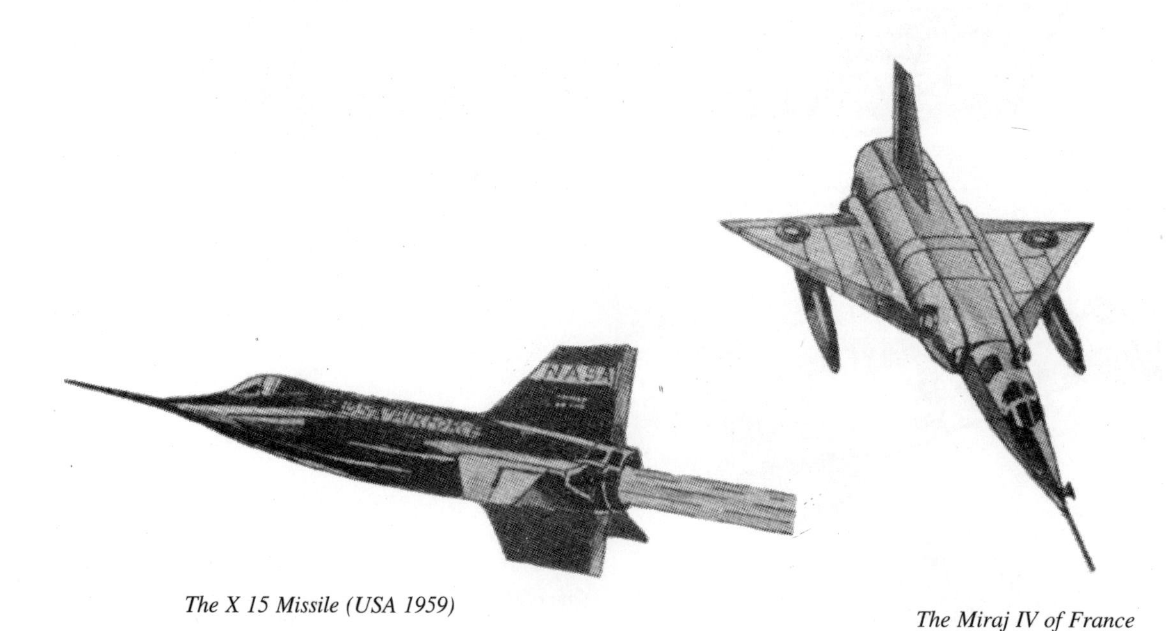

The X 15 Missile (USA 1959)

The Miraj IV of France

JET PLANE

A jet plane works on the principle of Newton's third law of motion. This law states that for every action there is an equal and opposite reaction. In a jet engine, combustion of fuel is an action and exhaust of the gases from the nozzle is the reaction. The gases move at high speed and, as a result, the engine moves in the opposite direction of the exhaust gases. Jet engines are 80% more efficient as compared to piston engines.

A jet engine, while flying, leaves a white line of smoke. A jet plane makes use of jet propulsion. A liquid or gas when it comes out of a small hole is called a jet. When a jet comes out with a great speed, its reaction makes the thing move in the opposite direction of the jet. The jet engine of airplanes has a gas turbine in the rear portion and a compressor fan in the front. A jet engine uses oxygen in the air around it.

The idea of employing jet engines in airplanes was proposed for the first time by Captain Marcone of France, Henry Coanda of Rumania and Maxime Gillaume in 1921. In 1930, Frank Whittle got a patent of a jet-propelled airplane. The first experimental flight of a jet-propelled airplane was proposed by Frank Whittle, and constructed by the British Power Jets Ltd. which took place on April 12, 1937. The first flight of a turbojet-propelled airplane was materialised by Heinkel at Marien He in Germany on August 27, 1939. After this, different types of gas turbine-propelled engines came into existence.

A jet engine can fly at an altitude of 3,05,000 metres. Their speeds go higher than 3,500 km/hr. Jet planes have added a new chapter in aeroplane technology and have made the world smaller in size. Concord is a world-famous jet plane. Lockheed jet plane had broken the world record in 1976 by flying at a whopping speed of 3,529.56 km/hr. Columbia flew at a magnificent speed of 26,715 km/hr. Boeing 707 is also a world-famous aeroplane.

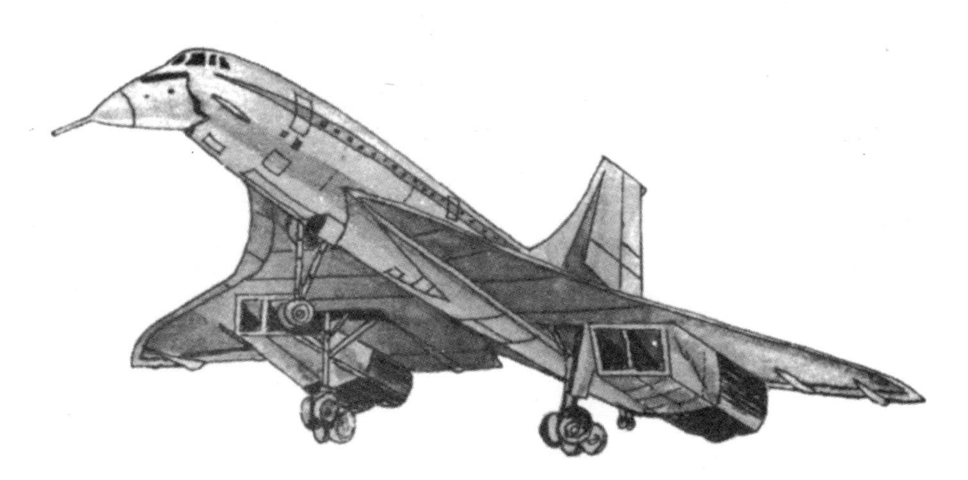

The Anglo-French Supersonic Cancord (1969)

HELICOPTER

A helicopter is an aircraft which can fly straight up or down in air without any runway. It can hover in one position like a humming bird. A helicopter can fly in any direction. It can fly upwards or downwards and even forward or backward.

Helicopter is a Greek word made up of 'heli' and 'copter', which means screw and fan. It was known to the Chinese about 500 BC as a flying toy. In about 1500, Leonardo Da Vinci made a design of a helicopter, based on Archimedes nut and screw principle. During the 19th century, people tried to develop helicopters but only in 1936 Focke-Wulf Company of Germany succeeded in developing a helicopter. This first helicopter in 1937 could fly at a height of 11,000 ft. reaching a speed of 70 miles per hour. After this Igor Sikorsky, a Russian engineer, made his first helicopter for the United States Army, which was named as XR-4. It was successfully tested in 1941. Since then many designs of helicopters have appeared in the world.

Nowadays, helicopters are used to evacuate sick persons from the sea, flood victims, and delivering food, medicines and other supplies to the army and stranded communities. These are also used to spray insecticides on crops and for survey. Every off-shore oil rig has its helicopter platform. They check remote power lines, monitor forest fires, hunt submarines and find a shipping path through ice-field. Helicopters with small tanks are used by army also. The uses of helicopters are multiplying day by day. The helicopters used in wars are of modern technology.

A Helicopter

SPACE SHUTTLE

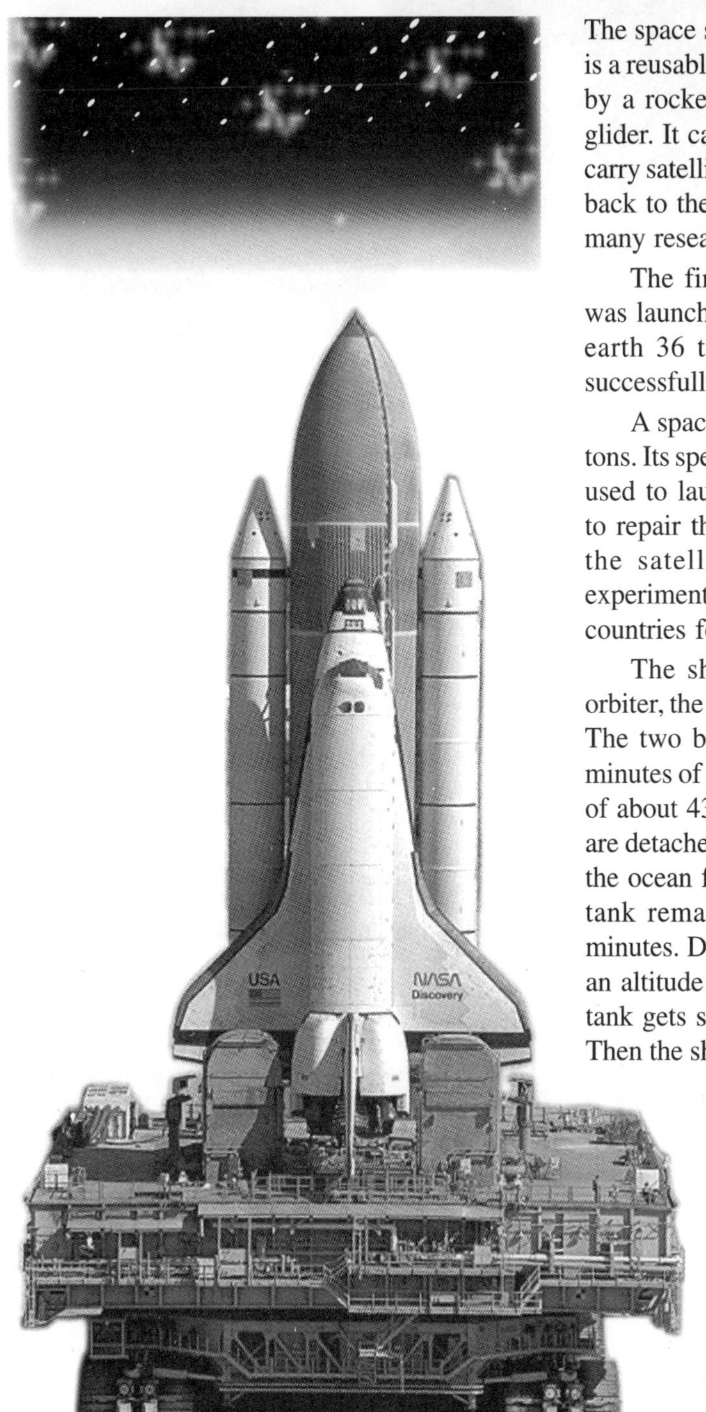

The space shuttle, developed by U.S. scientists, is a reusable spacecraft. It is launched into space by a rocket but it returns to earth like a huge glider. It can be reused for space travels. It can carry satellites into orbit and even bring satellites back to the earth for repair. It is being used in many researches in space.

The first space shuttle, named Columbia, was launched on April 12, 1981. It orbited the earth 36 times and came back to the earth successfully on April 14, 1981.

A space shuttle can carry a load of about 21 tons. Its speed is about 28,000 km/hr. It is mainly used to launch satellites and space telescopes, to repair the defective satellites, to bring back the satellites and to carry out scientific experiments in space. It can be hired by other countries for scientific missions.

The shuttle has mainly three parts—the orbiter, the external tank and the booster rockets. The two booster rockets give the shuttle two minutes of sturdy thrust and take it to an altitude of about 43 kms. After this the booster rockets are detached from the shuttle and parachute into the ocean for recovery and reuse. The external tank remains with the shuttle for about ten minutes. During this period the shuttle acquires an altitude of 200 kms. After this, the external tank gets separated and falls back to the earth. Then the shuttle is taken into its orbit by its two

The launching of space shuttle

Engines are being started for landing

Orbiter enters into its orbit

Orbiter enters into space

Space shuttle has landed on the earth.

External tank being separated out

Booster rockets being separated out

engines. It continues to orbit the earth as long as it is desired. The engines are used again at the end of the mission to bring the orbiter back into the atmosphere. After re-entry, the orbiter glides down through the atmosphere and lands like an aircraft on a long run-way. Apart from the large fuel tank, all other parts of a space shuttle can be reused.

In addition to Columbia, the Discovery, Challenger and Atlantis space shuttles have made several flights. On January 26, 1986, the Challenger exploded just after the take off, killing all the crew members on board. This has been one of the biggest setbacks in the history of the space shuttle. Another space shuttle named Columbia exploded in the year 2003 (1st February, 2003) shortly before landing on the earth. Kalpana Chawla of the Indian origin and all other members of the crew died in this fatal accident. ■■